Vedic Mathematics

Vedic Mathematics

A HANDBOOK

INDIRA SUKUMARAN

PGT MATHEMATICS & SUPERVISOR

INDIAN SCHOOL SOHAR, SULTANATE OF OMAN

AUTHORSPRESS

Worldwide Circulation through Authorspress Global Network
First Published in 2021
by
Authorspress
Q-2A Hauz Khas Enclave, New Delhi-110 016
Phone: (0) 9818049852
E-mail: authorspressgroup@gmail.com
Website: www.authorspressbooks.com

Vedic Mathematics: A Handbook
ISBN 978-93-90891-07-8

Printed in India at Thomson Press (India) Limited

This book is dedicated to my dear son Aditya,

Dear students and Teachers empowering!

- Indira Sukumaran

CONTENTS

FOREWORD

Indian culture dated back to the seventh century BC, is a treasure house of knowledge, manual and intellectual skills and techniques of human race. The accounts of these compiled in VEDAS (means treasure house of knowledge) exhibit the right and clear evidence.

It was Jagadguru Sankaracharya Sri Bharati Krishna Tirthaji of Govardhan Matt, Puri who had unearthed the real worth of mathematical techniques used by ancient Indians of Vedic period. He was an eminent scholar equally well-equipped and qualified not only in languages English and Sanskrit but in Mathematical and Physical Sciences too. He was a skilled writer and an eloquent orator. He studied the Suthras and Upasuthras in the appendix of Adharva Veda and developed several possible applications of these in Mathematics. Though he planned to write and bring out 16 volumes of all the branches of Mathematical computations and processes, only the introductory volume, with brief mention, yet of high merit, is left for reference. His disciples had compiled several notes on his lecture-demonstrations and speeches. Only these scripts are available now.

The Adharva Veda from among the four Vedas has hints in the form of Sutras and upasuthras in its Parishista which were used in mathematical computations.

It is indeed an excitement-provoking truth that the computations with numbers could be done so skillfully to reach the right answers in seconds just by tossing numbers in certain patterns! The conventional mathematics taught from First standard to Twelfth in the present times could be learnt in at the most two years by children in the age of 10 years!

There are several merits of the age old Vedic methods over the Western Conventional Mathematics presently learned world-wide.

Those are in brief:

- The aphorisms do apply and cover each and every branch of Mathematics.
- The Sutras are easy to understand, apply, remember, short and the computations can be done mentally in a very short time.
- There is no much complicated multi-step problem solving and the Vedic methods take a fraction of the time required compared to that for conventional modern mathematics.

Further, the Conventional methods activate the left part of the human brain and the right side which needs pattern based skill-exercises remain inert or slow in growth. Vedic methods, all of them involve pattern drilling and so training in these fields will ensure the growth of the entire brain.

It is certainly an advantage of teenage children to know the Vedic computations especially when they face the speed-based computational skill tests.

The students mastering these methods are sure to outsmart in problem solving.

Here, with extreme pleasure and pride I present my student Indira Sukumaran a very capable and skilled Mathematics teacher who has done a marvelous work by studying the 16 Sutras and 13 Upasutras in Veda Ganitham and has brought out a book on their applications with many illustrative examples. Each and every application is explained well supported by examples, in this book.

This handbook in Veda Ganitham is for sure an excellent guide to the beginners for quickly grasping the computational techniques, thereby to master over the high speed mathematical skills.

I congratulate Indira for her apt presentation of the work and wish to set higher achievements in her scholastic domain.

With love and regards,

Geetha Ravindran
Mathematics teacher (Retd),
Kendriya Vidyalaya

MESSAGE

Math is beautiful. But sometimes, this can be hard to see, and even harder to convey to students. Ms. Indira Sukumaran's *Vedic Mathematics - A Handbook* provides practical strategies and activities to help students, even those who are convinced that they are not good in mathematics.

During her teaching career spanning 22 years she has studied and followed various strategies to teach mathematics to thousands of students through middle and senior grades; to study how they learn and to find the most effective ways to unleash the math potential in all students

The book clearly and accurately

- Explains 16 sutras and 13 sub-sutras of Vedic mathematics
- Reveals how to solve complex problems in an easy and faster way
- Provides examples of rich mathematical activities.

I am convinced that this book will empower teachers and students.

Best wishes!

Mrs. Sanchita Verma
Principal, Indian School Sohar

ACKNOWLEDGEMENTS

My work is a collaborative product of many minds and their intelligence. I was inspired by a workshop on Vedic Mathematics way back in 2001, when I was teaching at Jamshedpur Public School, Jharkhand, India. That session aroused my interest and stimulated my mind towards the subject. I was keen to learn more about the subject in detail. Vedic Mathematics has its vibrant application in many topics of Mathematics. After studying it in detail, I wanted to jot down all the basic rules and simple applications in a form that could easily be accessed by the students and teachers, so that it can be used in our day today solving of problems in Mathematics.

I am indebted to Swami Sri Bharati Krsna Trithaji Maharaja for bringing out the book on Vedic Mathematics, which inspired many mathematicians, teachers and students to understand the various methods for solving problems mentally. While conceptualizing and writing this book, references are taken from many books and websites. All the reference is purely used for educational purpose. I am thankful to various books and websites on Vedic mathematics for helping me to bring out this handbook.

I thank my Mathematics teacher Mr. Vasu, who inspired me to study mathematics when I was a student studying class 10 at Kendriya Vidyalaya Avadi, Madras, which inspired me to become a Mathematics teacher today.

I extend my heartfelt thanks to Mrs. Sanchita Verma (Principal, Indian School Sohar, Oman), who has inspired and motivated me to do something novel. She is a true mentor for me in many ways. She is the co-author of the CBSE Handbook- *Pedagogical Leadership: A Handbook for Leading Learning in Schools.*

I extend my gratitude to my teacher Ms. Geeta Ravindran (Mathematics Teacher Retd. Kendriya Vidyalaya). I was fortunate to be her student during my school days. She is a very well-known poet and writer. She has published many books of different genres like short stories, poems, Vedic Maths in English and Malayalam. I am fortunate to avail her guidance to proceed with my first book. I am thankful to her, for her kind support and guidance.

I express my whole-hearted gratitude to the publishing house Authorspress for their guidance and support in bringing out this book.

I am ever grateful to my dear parents Smt. P. Kamalam and Shri. T. S. Nair. They showered their blessings and guided me in all my endeavors. I am also happy to receive all the encouragement, motivation, support from my son (Aditya Vinod), husband (Mr. K.Vinod), my brother (Mr. Hari. S), in bringing out this book.

I am thankful to the almighty for all the love, care, and blessing bestowed on us.

Indira Sukumaran

AUTHOR'S NOTE

Mathematics holds a prominent place in human life, because of its practical applications and implications. From the simple calculations that we do in our everyday life to the difficult calculations carried out in science and technology, Mathematics has an impact on and everything around us.

I have done my schooling from different Kendriya Vidyalayas spread across our country (Jodhpur, Delhi, Shillong, Madras, and Nasik) as my father was in Air Force. The guidance and power of knowledge acquired from my school teachers and parents inspired me to become a Mathematics teacher.

I completed my graduation with B.Ed. from Regional College of Education, Bhopal, and post-graduation from Pune University.

In my 22 years of Mathematics teaching in schools in India and abroad, I have observed that we can bring about remarkable changes in the way students perceive mathematics and their ability to succeed in the field by using different strategies of teaching and learning in Mathematics. There are various strategies that can be adopted: like Creativity in Mathematics, Investigating Mathematics, Group work, Recreational Mathematics, etc.

I feel Vedic Mathematics can be used as Recreational Mathematics or Investigating Mathematics strategy, along with the conventional method to generate more interest in the subject and also to erase the fear of the subject from the young mind.

Vedic Mathematics is the name given to the ancient system of Mathematics. Vedas are ancient holy texts from India and Vedic knowledge is in the form of shlokas or poems in Sanskrit verse. Vedic Maths refers to mathematical formulae called "sutras". It deals with techniques of calculations based on simple rules describing the natural ways of solving various mathematical problems.

Vedic Maths is a set of 16 Sutras (formulas) and 13 Sub-Sutras (corollaries) that were discovered from the Vedas by Swami Bharati Krishna TirthaJi. He developed mathematical principles around these Sutras and Sub-Sutras and published them in 1965 in a book named Vedic Mathematics.

Vedic Mathematics introduces the wonderful applications to addition, subtraction, multiplication, division, theory of numbers, compound multiplications, algebraic operations, factorizations, simple quadratic and higher-order equations, simultaneous quadratic equations, partial fractions, calculus, squaring, cubing, square root, cube root, and coordinate geometry, etc. In my opinion, Vedic Mathematics can be applied from class 6 to class 12 in the school system.

Vedic Mathematics has numerous advantages, to mention a few:

- Vedic Mathematics enriches knowledge and understanding of Mathematics.
- Vedic Mathematics methods come as a boon for all competitive exams (as mathematic problems can be solved with amazing accuracy and speed).

- Vedic Mathematics can be learned and used with ease if practiced regularly in a very short time.
- Vedic Mathematics system also provides a set of independent crosschecking methods.
- It sharpens the mind and increases mental agility and intelligence.
- It enhances the precision of thought.
- Vedic Math is complementary to regular Math taught in schools.

This book will be immensely useful for the teachers and students and will help to develop logical thinking and enhance confidence. This book presents a large number of problems and solutions to understand the application of Vedic Mathematics. The effectiveness of Vedic Maths cannot be fully appreciated without actually practicing the system.

I extend my sincere gratitude to Swami Bharati Krishna Tirtha Ji for his work on Vedic Mathematics and let us know the contents in Vedas in a simplified manner. Through this handbook, I am trying to reach out to students and teachers with the introduction of Vedic Mathematics, its basic application will be beneficial in the teaching and learning process.

Indira Sukumaran
M.Sc. B.Ed (Mathematics)
PGT Mathematics & Supervisor
Indian School Sohar
Sultanate of Oman

SUTRA 1
EKADHIKENA PURVENA

"By one more than the previous one"

APPLICATION -I

Squares of numbers ending in 5:

For any integer ending with 5, the square always ends with 25 and begins with the multiple of the previous integer and one more than the integer. For example:

45^2: Here the number is 45. We have to find out the square of the number.

For the number 45, the last digit is 5 and the 'previous' digit is 4. Hence, one more than the previous digit is 4+1=5.

Step 1: Multiply the previous digit 4 by one more than itself, that is, by 5 it becomes the L.H.S (left-hand side) of the result, that is, 4 X 5 = 20.

Step 2: The R.H.S (right-hand side) of the result is 5^2, that is, 25.

Step 3: 45^2 = 4x5/ 25 = 2025.

More examples:

65^2 = 6x7/25 =4225

125^2 = 12 x13 /25 = 15625

55^2 = 5x6 / 25 = 3025

15^2 = 1 x 2 /25 = 225

Note: It is applicable only for the numbers ending with 5

APPLICATION –II

Value of 1 / 19 (using division method)

The number of decimal places before repetition is the difference between numerator and denominator, i.e., 19 - 1=18 places.

For the denominator 19, the Purva means (previous) is 1.

Hence Ekadhikena Purva (one more than the previous) is 1 + 1 = 2.

The sutra is applied in a different context. Now the method of division is as follows:

1/19= 1/20 = **0. 0 5 2 6 3 1 5 7 8.... this is Quotient**

(this is the final answer)

1 0 1 0 0 1 1 1 1.....this gives Remainder

. 0 5 2 6 3 1 5 7 8 .. → Q

1 0 1 0 0 1 1 1 1 .. → R

Explanation is given as follows:

Step 1: Divide numerator 1 by 20.

i.e., 1 / 20 = 0.1 / 2 = .10 (0 times, 1 remainder)

Step 2: Divide 10 by 2

i.e., 0.05(5 times, 0 remainder)

Step 3: Divide 5 by 2

i.e., 0.0512 (2 times, 1 remainder)

Step 4: Divide 12 i.e., 12 by 2

i.e., 0.0526 (6 times, No remainder)

Step 5: Divide 6 by 2

i.e., 0.05263 (3 times, No remainder)

Step 6: Divide 3 by 2

i.e., 0.0526311(1 time, 1 remainder)

Step 7: Divide 11 i.e., 11 by 2

i.e., 0.05263115 (5 times, 1 remainder)

Step 8: Divide 15 i.e., 15 by 2

i.e., 0.052631517 (7 times, 1 remainder)

Step 9: Divide 17 i.e., 17 by 2

i.e., 0.05263157 18 (8 times, 1 remainder)

Step 10: Divide 18 i.e., 18 by 2

i.e., 0.0526315789 (9 times, No remainder)

Step 11: Divide 9 by 2

i.e., 0.0526315789 14 (4 times, 1 remainder)

Step 12: Divide 14 i.e., 14 by 2

i.e., 0.052631578947 (7 times, No remainder)

Step 13: Divide 7 by 2

i.e., 0.05263157894713 (3 times, 1 remainder)

Step 14: Divide 13 i.e., 13 by 2

i.e., 0.052631578947316 (6 times, 1 remainder)

Step 15: Divide 16 i.e., 16 by 2

i.e., 0.052631578947368 (8 times, No remainder)

Step 16: Divide 8 by 2

i.e., 0.0526315789473684 (4 times, No remainder)

Step 17: Divide 4 by 2

i.e., 0.05263157894736842 (2 times, No remainder)

Step 18: Divide 2 by 2

i.e., 0.052631578947368421 (1 time, No remainder)

Now from step 19, i.e., dividing 1 by 2, Step 2 to Step 18 repeats thus giving 1 / 19 =0.052631578947368421 or 0.052631578947368421

The conversion of a fraction (1/19, 1/29, 1/49 etc.) to its equivalent recurring decimal involves 18, 28 or 42 or more steps of cumbrous working (according to the conventional system). By using Sutras it requires only one single and simple step of mental working.

Applications of SUTRA 1:

- To find the square of numbers ending with 5.
- To find the decimal form of fractions.

SUTRA 2
NIKHILAM NAVATASCARAMAM DASATAH

"All from 9 and the last from 10"

The formula can be applied in the multiplication of numbers, whichare nearer to bases like 10, 100, 1000 i.e., to the powers of 10.

The difference between the number and the base is termed as deviation or deficiency. The deviation may be positive or negative. The positive deviation is written without the positive sign and the negative deviation is written using Rekhank (a bar on the number). Some examples are given in the following table.

NUMBER	BASE	NUMBER-BASE	DEVIATION
15	10	15-10	5
6	10	6-10	-4 OR $\overline{4}$
97	100	97-100	-03 OR $\overline{03}$
107	100	107-100	07
993	1000	993-1000	-007 OR $\overline{007}$
1014	1000	1014-1000	014

PROCEDURE:

- **Step 1**: Subtract the numbers with their closest multiple of 10 and multiply the results (deficiencies).
- **Step 2**: Subtract the results with other numbers.

- **Step 3**: Write the result of Step 2 in the beginning and the result of Step 1 at the end.

  For example: 95 x 99 =?

  100 – 95 = 5 and 100 – 99 = 1

  5 x 1 = 05

  95 – 1 or 99 – 5 = 94

  94 and 05

  So, the answer is 9405

  More examples:

BASE 10

(Deficiency from base)

9 X 7 = 63

9	-1
7	-3
6 /	**3**

9 -3 = 6
7-1 = 6
(-1) X (-3) = 3

8 X 6 = 48

8	-2
6	-4
4 /	**8**

8 -4 = 4
6-2 = 4
(-2) X (-4) = 8

BASE 10

12 X 13 = 156

12	+2
13	+3
15 /	**6**

12+3 = 15
13 + 3 = 15
2 x 3 = 6

13 X 7 = 91

13	+3
7	-3
10 /	**$\bar{9}$**

10 - 1 / 10-9

9 1

13 – 3 = 10
7 + 3 = 10
3 x (-3) = $\bar{9}$

SUB BASE 50 [10 X5 = 50]

48 X 49 = 2352

```
 48  -2
 49  -1
 ---------
 47 / 2
 x 5
 ---------
 2352
```

48-1= 47
49- 2 = 47
2 x 1 = 2

BASE 100

83 X 96 = 7968

```
 83  - 17
 96  - 0 4
 ---------
 79 / 68
 ---------
```

83-04=79
96 -17 = 79
(-17) x (-4) = 68

98 X 97 = 9506

```
 98  - 02
 97  - 03
 ---------
 95 / 06
 ---------
```

98-03= 95
97-02= 95
(-02) x(-03) = 06

BASE 100

107 X 96 = 10272

107 +7
96 - 4

$103 / \overline{28}$

103 -1 /100-28

10272

107-4=103
96 +7 = 103
$7 \times (-4) = \overline{28}$

106 X 99 = 10494

106 +6
99 - 1

$95 / \overline{06}$

105-1/ 100-06

10494

106-1= 105
99 +6= 105
$6 \times (-1) = \overline{06}$

BASE 1000

996 X 999 = 995004

```
 996   -04
 999   -01
 ----------
 995/ 004
```

996-01=995
999 -04= 995
(-4)x (-1)= 4

{Must be represented as a digit equal to the number of zero in the base or the highest base}

1003 x 996 = 998988

```
 1003   +3
  996   -4
 ----------
 999/ 012  (012 with bar)
 999-1/ 1000-012
 ----------
 998988
```

$999/\overline{012}$

1003-4=999
996+3 = 999
3 x (-4)= $\overline{012}$

BASE 10000

898 X 9996 = 8976408

```
 898    -9102
 9996   -4
 ------------
 894 /3 6408
 ------------
 894 + 3/ 6408
 ------------
 8976408
```

898 – 4 = 894
9996 - 9102 =894
9102 x 4 = 36408

BASE 10000

10005 x 9991 = 99959955

```
 10005   +5
 9991    -9
 ------------
 9996 /0045  (0045 with bar)
 ------------
 9996- 1/ 10000-45
 ------------
 99959955
```

$9996/\overline{0045}$

10005 – 9 = 9996
9991 + 5 = 9996
5 x -9 = $\overline{0045}$

COROLLARY

Squaring Technique

A. The number is smaller than the base or having a negative deficiency

Step 1: Decrease the given number by its deficiency from the base; it is the left-hand part of the answer

Step 2: the Square of the deficiency is the right-hand part of the answer.

e.g.: $\mathbf{(97)^2}$ Base= 100, Deficiency = -03

$= 97\text{-}3/\ (-3)^2$

= 9409

More examples:

$(987)^2 = (987-13)/(-13)^2$

$= 974169$

Base = 1000
Deficiency = -13
987 - 13 = 974
$(-13)^2$ = 169

OR

987 X 987 = 9506

987 - 13
987 - 13

974 / 169

987 - 13 = 974
987 - 13 = 974
(-13) x (-13) = 169

B. The number is greater than the base or having a positive deficiency

Step 1: Increase the given number by its deficiency from the base; it is the left-hand part of the answer

Step 2: the Square of the deficiency is the right-hand part of the answer.

e.g.: $(102)^2$ Base= 100, Deficiency = 02

$= 102+2/\ 2^2 = 10404$

More examples:

$(112)^2 = (112+12 / (12)^2$
$= 12544$

Base = 100
Deficiency = +12
112 + 12 = 124
$(12)^2 = 144$

OR

112 x 112 = 12544

```
 112  +12
 112  +12
 ---------
 124 / 144
 ---------
```

112 + 12 = 124
112 + 12 = 124
(12) x (12) = 144

$(10114)^2 = (10114 + 114) / (114)^2$
$= 10228 / 12996$
$= 102292996$

Base = 10000
Deficiency = +114
10114 + 114 = 10228
$(114)^2 = 114 + 14/(14)^2$
= 128/196
= 12996

OR

10114 x 10114 = 102292996

```
 10114   +114
 10114   +114
 -------------
 10228 / 12996
 -------------
   102292996
```

10114 + 114 = 10228
10114 + 114 = 10228
114 x 114 = 12996

$(49)^2 = (49-1) / (-1)^2$
$= 48x\ 5/1$
$= 2401$

Sub Base = 50 (5 x10)
Deficiency = -1

OR

49 X 49 = 2401

```
 49  - 1
 49  - 1
 --------
  48/1
  x5
 --------
  2401
```

49 - 1 = 48
49 – 1 = 48
(-1) x(-1) = 1

DIVISION (USING NIKHILAM METHOD)

Step 1: For two-digit numbers split each dividend into two parts by using a vertical line.

Step 2: The first part becomes the quotient and the second part obtained after the sum becomes the remainder

Example: 12 ÷ 9

i) 9 divides 1 zero times
ii) now, 1+0 = 1 (left side)
iii) add the left part with the right digit to get the right side

9) 1 / 2 (
   0/1
―――――
Q ← 1 / 3 → R

136÷9

	9√ 13 /6	
(9 x 1 = 9)	1/ 4	(13 -9 = 4)
(13 + 1 = 14)	14/ 10	(6+4 =10)
(14 + 1 = 15)	15/ 1	(10-9=1)
	Q R	

512 ÷ 9

	9 √5 1 /2	
9x 5 = 45	5 / 6	51 - 45= 6
	5 6 / 8	
	Q R	

Division by 9

Consider

312 ÷ 9

Add the first digit 3 to second digit 1 getting 3 + 1 = 4. Hence Quotient is 34. Now the second digit of 34 i.e., 4 is added to third digit 2 of dividend to get the remainder i.e., 4 + 2 = 6

2113 ÷ 9

i) Add first digit 2 to the second digit 1. 2 + 1 = 3

ii) Add the second digit of quotient 23. i.e., 3 to third digit '1' and obtain theQuotient. 3 + 1 = 4, 234

iii) Add the third digit of Quotient 234 i.e., 4 to the last digit '3' of the dividend, and write the final Quotient and Remainder. R = 4 + 3 = 7, Q = 234

Applications of SUTRA 2:

- To find the square of numbers near to base 10,100 etc.
- To find the product of two numbers near to the base.
- To divide any number by 9.

SUTRA 3
URDHVA – TIRYAGBHYAM

"Vertically and crosswise"

This Sutra is useful for the elementary multiplication of a large number by another large number.

For multiplication of any two two-digit numbers,

- Step 1: Multiply the last digits
- Step 2: Multiply numbers diagonally and add them.
- Step 3: Place Step 1 at the end and Step 2 at the beginning.
- Step 4: Multiply the first digit of both the number and put it at the beginning.
- Step 5: For the result, more than 2 or more digits, add the beginning digits to the beginning numbers.

Example:

32 x 45 =?

2x 5 = 10

(3 x5) + (2 x 4) = 15 + 08 = 23

3 x 4 = 12

32 x 45 = 12 | 23 | 10 = 12 | 23 + 1 | 0 = 12 | 24 | 0

= 12 + 2 | 40 = 1440

More examples:

111 X 112 = 12432

```
 1  11
 1  12
-----------
1 / 23/ 132
-----------
1/ 24/32  ⟶
-----------
 12432
```

Base = 100
11 X 12= 132
1 x 11+ 12 X 1 = 23
1 x 1 = 1
2 digits only allowed

1115 x 1004 = 1119460

```
  1 115
  1 004
  ---------
1 / 119/ 460  ⟶
  ---------
  1119460
```

Base = 1000
115 x 4 = 460
115 x 1+ 4x1=119
1 x 1 = 1
3 digits allowed

14 X 15 = 210

```
   14
  15
 -----
 1 / 9/ 20
 -----
 1/9+2/0
 1/11/0
1+1/1/0
 -------
 210
```

Base = 10
4 x 5= 20
1 x 5 + 4 x 1 = 9
1 x 1 = 1

27 x 32 = 864

```
   2 7
   3 2
  -----
6 / 25/ 14
  -----
6/25+1/4
6/ 26 /4
6+2 /6/4
 -------
 864
```

Base = 10
7 x 2 = 14
2 x 2 + 6 x 3 = 25
2 x 3 = 6

Application of SUTRA 3:

- To find the product of any two numbers.

SUTRA 4
PARAVARTYA YOJAYET

'Transpose and Adjust'

This Sutra can be easily applied when we consider the division by divisors of more than one digit, and when the divisors are slightly greater than powers of 10.

Example 1: Divide 1145 by 11.

Step 1 : (From left to right) write the Divisor leaving the first digit, write the other digit or digits using negative (-) sign and place them below the divisor as shown.

11

-1

Step 2: Write down the dividend to the right. Set apart the last digit for the remainder.

i.e.,	11	114	5
	- 1		

Step 3: Write the 1st digit below the horizontal line drawn under the dividend. Multiply the digit by (–1), write the product below the 2nd digit and add.

i.e.,	11	114	5
	-1	-1	
	10		

Since 1 x –1 = -1and 1 + (-1) = 0

Step 4: We get second digits' sum as '0'. Multiply the second digits' sum thus obtained by (–1) and write the product under 3rd digit and add.

11	114	5
- 1	-10	
	104	5

Step 5: Continue the process to the last digit.ie -1x 4 = 4 write it below the last digit 5

i.e.,	11	114	5
	- 1	-10	-4
		104	1

Step 6: The sum of the last digit is the Remainder and the result to its left isQuotient. **Thus Q = 104 and R = 1**

Examples:

2446 ÷ 12

12	244	6
- 2	- 40	- 8
	204	- 2
	203	12- 2
Q ←	203	10 → R

STEPS

1) -2 x 2= 4
2) 4-4 =0
3) 4 + 0=4
4) 4 x (-2) = -8
5) 6 – 8 = -2

432 ÷ 11

11	4 3	2
-1	-4	1
	Q ← 3 9	3 → R

OR

11	4 3	2
$\bar{1}$	$\bar{4}$	1
	4 $\bar{1}$	3
	40-1 /	3
	Q ← 39 /	3 → R

STEPS

1) -1 x 4 = -4
2) 43 - 4 = 39 (Q)
3) -4 + 3 = -1
4) (-1) x (-1) = 1
5) 2 + 1 = 3 (R)

546 ÷ 11

11	5 4	6
-1	-5	1
	Q ← 4 9	7 → R

OR

11	5 4	6
$\bar{1}$	$\bar{5}$	1
	5 $\bar{1}$	7
	50-1 /	7
	Q ← 49 /	7 → R

STEPS

1) -1 x 5 = -4
2) 54 - 5 = 49 (Q)
3) 4 – 5 = - 1
4) (-1) x (-1) = 1
5) 6 + 1 = 7 (R)

Division can be done by using the Nikhilam method here. Split each dividend into two parts for Q and R. The number of digits on the right side of the slash must be the same as the number of digits in the complement of the divisor.

Example: Compliment of 89 = 11 (by using the Nikhilam rule about subtraction of all from 9 and of the last from 10)

DIVISION USING NIKHILAM METHOD

12 ÷ 9   9) 1 / 2
Compliment 1   0 / 1   1 x 1 = 1
Q ← 1 / 3 → R

33 ÷ 9   9) 3 / 3
Compliment 1   0 / 3   1 x 3 = 1
Q ← 3 / 6 → R

103 ÷ 9   9) 1 0 / 3
Compliment 1   1 / 1   1 x 1 = 1 ; 1+ 0= 1 x 1 =1
Q ← 1 1 / 4 → R

2 2 5 ÷ 9   9) 2 2 / 5
Compliment 1   2 / 4   1 x 2 = 2 ; 2 + 2 = 4 ; 4 x 1 = 4
2 4 / 9 ≥ 9
Q ← 2 5 / 0 → R

3 5 6 ÷ 9   9) 3 5 / 6
Compliment 1   3 / 8   1 x 8 = 8
3 8 / 14 ≥ 9   { 14 – 9 = 5(R) }
Q ← 3 9 / 5 → R

136 ÷ 9   9) 1 3 / 6
Compliment 1   1 / 4   1 x 1 = 1 ; 1+ 3= 4 x 1 =4
1 4 / 10 ≥ 9   { 10 – 9 = 1 (R) }
Q ← 1 5 / 1 → R

23 ÷ 8   8) 2 / 3

Compliment 2   0 / 2 x 2=4   2 x 2 = 4 ; 3 + 4 = 7(R)

Q   2 / 7   R

28 ÷ 8   8) 2 / 8

Compliment 2   0 / 2 x 2=4   2 x 2 = 4 ; 3 + 4 = 7(R)

2 / 12 ≥ 8   12 – 8 = 4

Q   3 / 4   R

12 ÷ 7   7) 1 / 2

Compliment 3   0 / 1 x 3=3   1 x 3 = 3 ; 3 + 2 = 5(R)

Q   1 / 5   R

111 ÷ 89   89) 1 / 11

Compliment 11   / 11   1 x 11 = 11

Q   1 / 22   R

12345 ÷ 8888   8888) 1 / 2345

Compliment 1112   / 1112   1 x 1112 = 1112

Q   1 / 3457   R

210012 ÷ 8997   8997) 21 / 0012

Compliment 1003   2 / 006   2 x 1003 = 2006

/ 3009   2 + 1 = 3 X 1003 = 3009

Q   23 / 3081   R

Application of SUTRA 4:

- Division: To find the quotient and remainder.

SUTRA 5
SUNYAM SAMYASAMUCCAYE

"When the sum is the same, the sum is equal to zero"

The term 'Samuccaya'has several meanings under different contexts.

i) We interpret, 'Samuccaya' as a term that occurs as a common factor in allthe terms concerned and proceeds as follows.

Example 1: The equation $4x + 9x = 2x + 7x$ has the same factor 'x 'in all its terms. Hence by the sutra it is zero, i.e., $x = 0$.

Otherwise, we have to work like this:

$4x + 9x = 2x+7x$

$13x = 9x$

$13x – 9x = 0$

$4x = 0; x=0$

E.g. $2x + 5x = 3x + x; x =0$

Example 2: $7(x-3) = 5(x-3)$

No need to proceed in the usual procedure like,

$7x -21 = 5x -15$

$7x – 5x = 21 – 15$

$2x = 6$ or $x = 6 \div 2 = 3$

Now, Samuccaya is (x -3); x - 3 = 0 gives x = 3

E.g. 2 (x+5) = 9 (x+5); x= -5

Example 3: (x -3) (x -6) = (x +9) (x + 2)

Here Samuccaya is -3 x -6 = 18 = 9 x 2 since it is same, we derive x = 0

E.g. (x +4) (x +6) = (x +2) (x+ 12); x =0

Example 4:

$$\frac{5x + 2}{2x + 3} = \frac{2x + 3}{5x + 2}$$

Since N1 + N2 = 5x + 2 + 2x + 3 = 7x+5, (N1 and N2 are the numerators)

D1 + D2 = 2x + 3 + 5x + 2 = 7x +5, (D1 and D2 are the denominators)

We have, N1 + N2 = D1 + D2 = 7x + 5

Hence from Sunya Samuccaya we get 7x + 5 = 0; x = -5/7

Example:

$$\frac{x - 7}{x + 3} = \frac{x + 3}{x - 7}$$

X-7 + x +3 =0; x -4 = 0; x =4

Application of SUTRA 5:

This Sutra helps to solve linear equations and quadratic equation, under specific conditions.

SUTRA 6
ANURUPYE – SUNYAMANYAT

"If one is in ratio the other one is 0"

We use this Sutra in solving a special type of simultaneous simple equations. In which the coefficients of 'one' variable are in the same ratio to each other as the independent terms are to each other. In such a context the Sutra says the 'other' variable is zero from which we get two simple equations in the first variable. On solving, both the equations give the same value for the variable.

Example 1: **solve** $2x + 3y = 4$ $3x + 9y = 12$

Observe that the y-coefficients are in the ratio 1: 3, which is same as the ratio of independent terms i.e., 4: 12 i.e., 1: 3. Hence the othervariable $x = 0$ and $3y = 4$ gives $y = 4/3$

E.g.$5x - 2y = 3$; $10x - 8y = 6$; x coefficients and constants are in the ration 1:2. So $y=0$; $5x=3$; $x=3/5$

Example 2: Solve $213x + 362y = 639$: $426x + 345y = 1278$

Here coefficient of x are in the ratio 213: 426 = 1: 2 and constant terms ratio is 639: 1278= 1: 2.

$y = 0$ and $213x = 639$ or $426x = 1278$ gives $x = 3$.

Application of SUTRA 6:

- This Sutra helps to solve simultaneous linear equations, under specific conditions.

SUTRA 7
SANKALANA – VYAVAKALANABHYAM

"By addition and by subtraction"

i) In two general equation such as ax + by = p and cx + d y = q, where x and y are unknown values, then by solving the simultaneous equation we can get the values of x and y (cross multiplication, elimination or substitution method)

x = (b q – p d) / (b c – ad)

y = (c p – a q) / (b c – ad)

We can use the above formula directly;

For example:

5x + y = 4 and 6x + y = 3

x = (3-4)/ (6-5) = -1

y = (15-24)/ (6-5) = -9

ii) It can be applied in solving a special type of simultaneous equations where the x - coefficients and the y- coefficients are found interchanged.

Example:

63x – 37y = 120

37x – 63y = 80

Add them,

i.e., $(63x - 37y) + (37x - 63y) = 120+80$ i.e., $100x - 100y = 200$

$x - y = 2$

Subtract one from other,

i.e., $(63x - 37y) - (37x - 63y) = 120-80$ i.e., $26x + 26y = 40$

$x + y = 40/26$ and repeat the same sutra,

We get $x = -3/13$ and $y = 23/13$

Application of SUTRA 7:

- This Sutra helps to solve simultaneous linear equations, under specific conditions.

SUTRA 8
PURANAPURANABHYAM

"By the completion or non-completion"

It can be applied to solve and find the roots of the equation.

Example 1: $x^3 + 9x^2 + 23x + 15 = 0$........ (1)

We know $(x + 3)^3 = x^3 + 9x^2 + 27x + 27$

So adding on the both sides of (1), the term (4x+12),

We get: $x^3 + 9x^2 + 27x + 27 = 4x+12$

$(x + 3)^3 = 4(x+3)$

$y^3 = 4y$

$y (y^2 - 4) = 0$; y=0, y = 2, y = -2

Therefore, x= -3, -1,-5

More examples:

$x^2 + 2x - 3 = 0$
$x^2 + 2x - 3 + 1 - 1 = 0$
$(x + 1)^2 - 4 = 0$ (as $(x + 1)^2 = x^2 + 2x + 1$)
$(x + 1)^2 = 4$
x +1 = 2, x +1 = -2
x =1; x = -3

$x^3 + 6x^2 + 10x + 4 = 0$

$x^3 + 6x^2 = -(10x + 4)$

$x^3 + 6x^2 + 12x + 8 = -10x - 4 + 12x + 8$

$(x + 2)^3 = 2x + 4$ (as $x^3 + 6x^2 + 12x + 8 = (x + 2)^3$)

$(x + 2)^3 = 2(x+2)$

$y^3 = 2y$

$y(y^2 - 2) = 0$

$Y=0 ; y = \sqrt{2} ; y = -\sqrt{2}$

$X= 2; x = \sqrt{2} - 2 ; x = -\sqrt{2} - 2$

Application of SUTRA 8:

- This Sutra helps to find the roots of an equation.

SUTRA 9
CALANA – KALANABHYAM

"Differences and Similarities"

1. In the first instance it is used to find the roots of a quadratic equation

For e.g.: $3x^2 + 2x - 5 = 0$.

$3x^2 + 5x - 3x - 5 = 0$

(X-1) (3x+5)=0

x= 1, x = -5/3

2. Another context

$\frac{AC+D}{BC+E} = \frac{A}{B}$, **then** $\frac{A}{B} = \frac{D}{E}$

$$\frac{x^2 + x + 1}{x^2 + 3x + 3} = \frac{x + 1}{x + 3}$$

$$\frac{x(x + 1) + 1}{x(x + 3) + 3} = \frac{x + 1}{x + 3}$$

Therefore, $\frac{x+1}{x+3} = \frac{1}{3}$**; solving x= 0**

3. It is used in differential calculus also.

Application of SUTRA 9:

- This Sutra helps to find the roots of an equation.

SUTRA 10
YAVADUNAM

"Whatever the extent of its deficiency"

It can be used to find the square of a number close to the multiple of 10.

Step 1: Find the deficiency of the given number from its base.

Step 2: consider, Number (+ or -) Deficiency which will be the Left-hand side of the result

Step 3: Consider the square of the deficiency (express it as a number having digits equal to the no. of zeroes in the base), which will be the right side of the result.

Example:

NUMBER	STEPS	DEFICEINCY
$(8)^2$	$(8-2)/(-2)^2$ = 64	8-10=-2
$(9)^2$	$(9-1)/(-1)^2$ = 81	9-10=-1
$(16)^2$	$(16+6)/6^2$ = 22/36 = (22+3)/6= 256	16-10=6
$(14)^2$	$(14+4)/4^2$ = 18/16= 18+1 /6= 196	14-10=4
$(98)^2$	$(98-2)/(-2)^2$ =96/04= 9604	98-100=-2
$(103)^2$	$(103+3)/3^2$	103-100=3

SUTRA 12
SESANYANKENA CARAMENA

"The remainders by the last digit"

We can find the decimal form of rational numbers by using this sutra. Some specific examples are given here:

Example: 1/13 = 0.076923

1. Find successive remainders, they are 1, 10, 9, 12, 3 and 4.
2. Write the first remainder at the end and proceed the remaining as in step 1 we get it as 10, 9, 12, 3, 4 and 1.
3. Multiply with the last digit of the divisor i.e. by (3): 30, 27, 36, 9, 12 and 3.
4. Now consider just last digits, that gives final answer: 0.076923

Example: 1/7 = 0.142857

1. As seen earlier find the successive remainders, they are 1, 3, 2, 6, 4 and 5.
2. We will write them as 3, 2, 6, 4, 5 and 1.
3. Multiply them with the last digit of divisor (7): 21, 14, 42, 28, 35 and 7
4. Now take their last digits and that's the final answer: 0.142857

SUTRA 13
SOPANTYADVAYAMANTYAM

"The ultimate and twice the penultimate"

1. For the equation in the format $\frac{1}{AB}+\frac{1}{AC}=\frac{1}{AD}+\frac{1}{BC}$, where A, B, C and D are in AP, the result is 2C (penultimate) + D (ultimate) = 0.

   Example:

   $$\frac{1}{(x+1)(x+3)}+\frac{1}{(x+1)(x+5)}=\frac{1}{(x+1)(x+7)}+\frac{1}{(x+3)(x+5)}$$

   Applying the formula, 2(x+5) + (x+7) = 0; x = -17/3

2. Antyayoreva

   $\frac{AC+D}{BC+E}=\frac{A}{B}$, then $\frac{A}{B}=\frac{D}{E}$

   $$\frac{x^2+x+1}{x^2+3x+3}=\frac{x+1}{x+3}$$

   $$\frac{x(x+1)+1}{x(x+3)+3}=\frac{x+1}{x+3}$$

   Therefore, $\frac{x+1}{x+3}=\frac{1}{3}$; solving x= 0

   $$\frac{1}{(x+a)(x+b)}+\frac{1}{(x+b)(x+c)}+\frac{1}{(x+c)(x+d)}$$

   $$=\frac{3}{(x+a)(x+d)}$$

   Here a, b, c and d are in AP

3. Sum of the series is a fraction whose numerator is the sum of the numerators in the series and whose denominators is the product of the 2 ends (binomials).

   E.g. $\frac{1}{(x+6)(x+7)}+\frac{1}{(x+7)(x+8)}+\frac{1}{(x+8)(x+9)}=\frac{3}{(x+6)(x+9)}$

Application of SUTRA 13:

- This Sutra helps to find the solution of specific equations.

SUTRA 14
EKANYUNENA PURVENA

"By one less than the previous one"

MULTIPLICATION WITH 9, 99 ...

I) Digits in the multiplicand and the multiplier (9) are same

Step 1: Left hand side = multiplicand – 1

Step 2: Right hand side = 9 – no. obtained in step 1

Example: 2 x 9 = (2-1)/ (9-1) = 18

3 x 9 = (3-1)/ (9-2) = 27

7 x 9 = (7-1)/ (9-6) = 63

14 x 99= (14-1) / (99-13) = 1386

233 x 999 = (233-1)/ (999-232) = 232767

5342 x 9999= (5342-1)/ (9999-5341) = 53414658

23531 x 99999 = (23531 -1) / (99999 – 23530) = 2353076469

222333444 x 999999999 = (222333444-1) / (999999999 – 222333443)

= 222333443777666556

II) **Multiplier (9) is multiplied with a two digit number**

Step 1: Left hand side = If the multiplicand begin with 1 then subtract no. by 2, if it begins with 2 then subtract 3, so on

Step 2: consider the compliment of number in the unit place of the multiplicand.

11 x 9 = 11-2 / 9 =99

14 x 9 = 14 -2/ 6 = 126

29 x 9 = 29-3 / 1= 261

36 x 9 = 36 -4 /4 = 324

57 x 9 = 57-6 / 3= 513

72 x 9 = 72-8 / 8= 648

98 x 9 = 98 – 10 / 2 = 882

III) **Multiplier has digits more than the multiplicand**

Step 1: left hand side = Multiplicand – 1

Step 2: Right hand side =multiplier – no. obtained in step 1

1 x 99 = 1-1/ 99-0 = 099

4 x 99 = 4 – 1/ 99 – 3 = 396

10 x 99 = 10-1 / 99 -9 = 990

17 x 999 = 17 – 1 / 999 -16 = 16983

Application of SUTRA 14:

- This Sutra helps to find the product of a number with 9, 99, 999 etc.

SUTRA 15
GUNITASAMUCCAYAH

"The product of the sum is equal to the sum of the product"

It is applied in quadratic equations, cubic, bi-quadratics etc.:

In order to verify the result, the product of the sum of the coefficients of in the factors is equal to the sum of the coefficients in the product.

Examples:

$(x + 8)(x + 5) = x^2 + 13x + 40$

Or, $(1+8)(1+5) = 1 + 13 + 40$

Or, $9x6 = 54$;

Or, $54 = 54$; thus verified.

$(x - 2)(2x + 1) = 2x^2 - 3x - 2$

Or, $(1 - 2)(2 + 1) = 2 - 3 - 2$

Or, $-1 \times 3 = -3$

Or, $-3 = -3$

$(x - 3)(x - 5) = x^2 - 8x + 15$

Or, $(1 - 3)(1 - 5) = 1 - 8 + 15$

Or, $-2 \times -4 = 8$

Or, 8 = 8

$(x+1)(x+2)(x+3) = x^3 + 6x^2 + 11x + 6$

Or, (1+1) (1+2) (1+3) = 1+6+11+6

Or, 2x 3 x 4 = 24

Or, 24= 24

$(x-1)(x-4)(x-5) = x^3 - 10x^2 + 29x - 20$

Or, (1-1) (1-4) (1-5) = 1-10+29-20

Or, 0 x -3 x -4 = 0

Or, 0= 0

Thus, if and when some factors are known, this rule helps us to fill in the gaps.

Application of SUTRA 15:

- This Sutra helps to find the factors of a given polynomial.

SUTRA 16
GUNAKASAMUCCAYAH

"The factors of the sum is equal to the sum of the factors"

For the quadratic equation, the factor of the sum of the coefficients of 'x' in the product is equal to the sum of the coefficients of 'x' in the factors.

Examples:

$(x + 2)\ (2x + 1) = 2x^2 + 5x + 2$

Or, $(1 + 2)\ (2 + 1) = 2 + 5 + 2$

Or, $3 \times 3 = 9$

Or, $9 = 9$

a) Here factors are -2 & -1/2; for any quadratic equation $ax^2 + bx + c = 0$

   Sum of the roots = -b/a

   i.e.; $-2 + (-1/2) = -5/2$

   $-5/2 = -5/2$

   Product of the roots = c/a

   i.e.; $-2 \times -1/2 = 2/2$

   $1 = 1$

b) Also for $2x^2 + 5x + 2 = 0$

We can find x by using, 2ax +b = $\pm\sqrt{b^2 - 4ac}$

We get x = -2, -1/2

(x - 7) (x - 3) = x^2 -10x + 21

Or, (1 - 7) (1- 3) = 1 - 10 +21

Or, -6 x -2= 12

Or, 12 = 12

a) Here, factors are 7 & 3; for any quadratic equation:

   $ax^2 + b x + c = 0$

   Sum of the roots = -b/a

   i.e.; 7 + 3 = - (-10)/1

   10 = 10

   Product of the roots = c/a

   i.e.; 7x3= 21/1

   2 1= 21

b) Also for x^2 -10x + 21=0

   We can find x by using, 2ax +b = $\pm\sqrt{b^2 - 4ac}$

   We get x = 7, 3

Application of SUTRA 16:

- This Sutra helps to find the roots of an equation.

SUB SUTRA 1
ANURUPYENA

"Proportionately"

This Sutra is useful for finding the products of two numbers when both of them are near the Common bases i.e. powers of base 10. For example:

i) 48 x 45

We take 50 as a base point because it is the nearest powers of base 10.

48 = 50 – 2

45 = 50 – 5

Multiplying the difference from 50 to the numbers,

2 x 5 = 10 ...... (a)

From cross subtraction,

48 – 5 = 43 or 45 – 2 = 43....... (b)

Adding (a) with the base i.e., 10+50 = 60 (goes on right hand side)

Dividing 43 by 2 give 21 with remainder 1 (goes on left hand side)

Thus, 48 x 45 = 2160 (by division method)

Or, consider 48 x 45 = 43 x 5 / 10 = 215/10 = 2160 (by multiplication method)

More examples:

51 X 57 = 2907

51 +1
57 +7
----
58 / 7
x 5
----
290 /7 ⟶ 2907

SUB BASE = 50
51 +7 = 58
57+1 = 58
1 X 7 =7
58 X 5 =290

81 X 83 = 6723

81 + 1
83 +3
----
84 / 3
x 8
----
672 /3 ⟶ 6723

SUB BASE = 80
81 +3 = 84
83+1 = 84
1 X 3 =3
84 X 8 =672

32 X 34 = 1088

32 + 2
34 +4
----
36 / 8
x 3
----
108 /8 ⟶ 1088

SUB BASE = 30
32 + 4 = 36
34 + 2 = 36
2 X 4 = 8
36 X 3= 108

47 X 54 = 2538

47 - 3
54 + 4
----
51 / $\overline{12}$
x 5
----
255 / $\overline{12}$

253/ 20 -12 = 2538

SUB BASE = 50
54 -3 = 51
47 +4= 51
-3 X 4 = - 12
51 X 5 =255

SUB SUTRA 2
SISYATE SESASAMJNAH

"Remainder remains constant"

Consider division of 65 successively by 2,4,8,16,32 and 64.

65/2=32 ½, 65/4= 16 ¼, 65/8 = 8 1/8, 65/16= 4 1/16, so on. We see that as the denominator i.e. the divisor goes on increasing in a certain ratio, the quotient goes on decreasing, proportionately; but the remainder remains constant. The same principle is applied in the following applications.

APPLICATION:

For multiplication of any numbers near to more than multiple of 10

Step 1: Subtract the closest multiple of 10 from the number and multiply the results.

Step 2: Add the results with other numbers.

Step 3: Write the result of Step 2 in the beginning and result of Step 1 at the end.

For example:

1) 102 x 105

102 – 100 = 2 and 105 – 100 = 5

2 x 5 = 10

102 + 5 or 105 + 2 = 107

107 and 10

So, the answer is 10710

Or,

102 +2 {Base = 100, 102+5= 105 + 2= 107, 2x 5 = 10}

105 +5

107 / 10

102 x 105 = 10710

2) 1003 x 1008

1003 – 1000 = 3 and 1008 – 1000 = 8

3 x 8 = 24

1003 + 8 or 1008 + 3 = 1011

1011 and 024

So, the answer is 1011024

Or, 1003 +3 {Base = 1000, 1003 + 8 or 1008 + 3 = 1011, 3 x 8 = 24}

1008 +8

1011 / 024

1003 x 1008 =1011024

SUB SUTRA 3
ADYAMADYE NAANTYAMANTYENA

"The first by the first and the last by the last"

Consider example:

$4x^2 + 12x + 5 = (2x + 1)\ (?)$

If we know one factor of the equation then we can find the other factor by using this Sutra.

Consider, $\left[\frac{4}{2}x + \frac{5}{1}\right] = 2x + 5$

I.e. divide the first coefficient of x^2 i.e. 4 by the coefficient of x of factor i.e. 2 then divide the last term of equation i.e. 5 by the last term of the known factor i.e. 1

We get, $4x^2 + 12x + 5 = (2x + 1)(2x+5)$

$6x^2 + 11x - 10 = (2x + 5)\ (?)$

If we know one factor of the equation then we can find the other factor by using this Sutra.

Consider, $\left[\frac{6}{2}x + \frac{-10}{5}\right] = 3x - 2$

I.e. divide the first coefficient of x^2 i.e. 6 by the coefficient of x of factor i.e. 2 then divide the last term of equation i.e. (-10) by the last term of the known factor i.e. 5

We get, $6x^2 + 11x - 10 = (2x + 5)(3x - 2)$

SUB SUTRA 4
KEVALAIH SAPTAKAM GUNYAT

For 7 the Multiplicand is 143 (Kevala: 143, Sapta: 7).

On the basis of 1/7, without any multiplication, we can calculate 2/7, 3/7, 4/7, 5/7 and 6/7. For that 1/7=0.142857 is to be remembered. But since remembering 0.142857 is difficult we remember Kevala (143).

1/7 = 0.142857

By using142857 we can get the result of the following by simple multiplication:

- 2/7 = 0.285714
- 3/7 = 0.428571
- 4/7 = 0.571428
- 5/7 = 0.714285
- 6/7 = 0.857142

All the values are in cyclic order.

So if we can remember 1/7 then we can obtain 2/7, 3/7 and etc. as

- 2/7: the last digit of answer must be 4 (2x7=14)
- 3/7: the last digit of answer must be 1 (3x7=21)
- 4/7: the last digit of answer must be 8 (4x7=28)
- 5/7: the last digit of answer must be 5 (5x7=35)
- 6/7: the last digit of answer must be 2 (6x7=42)

SUB SUTRA 5
VESTANAM

"By osculation"

Ekadhika is a test for divisibility and the process by which it serves this purpose is technically called vestana or osculation.

Value of 1 / 19

For the denominator 19, the purva means (previous) is 1.

Hence Ekadhikena purva (one more than the previous) is 1 + 1 = 2(osculation)

Now we use 2 for division purpose.

1/19= 1/20 = 0. 0 5 2 6 3 1 5 7 8.... this is Q (which is final answer)

1 0 1 0 0 1 1 1 1.....this gives R

$1 \div 2; Q = 0, R = 1$

$10 \div 2; Q = 5, R = 0$

05$\div 2; Q = 2, R = 1$

$12 \div 2; Q = 6, R = 0$

06$\div 2; Q = 3, R = 1$ so on

The numbers of decimal places before repetition is the difference of numerator and denominator, i.e., 19 -1=18 places

1 / 19 =0.052631578947368421

Divisibility by 7

Ekadhika for 7 is derived from 7 x7 = 49 and therefore it is 5

To check whether 21 is divisible by 7

We multiply the last digit i.e. 1 by the Ekadhika i.e. 5 and add the product i.e. 1 x5 = 5 to the previous digit i.e. 2, so we get 5 + 2 =7. This process is called osculation. If the result of the osculation is the divisor itself or a repetition of the previous result we say it is divisible by 7. So 21 is divisible by 7.

For 14; 4 x5 +1 = 21; 1x 5 +2 = 7; 14 is divisible by 7

For 28; 8 x5 + 2 = 42; 2 x 5 + 4 = 14; 4 x5 + 1 = 21; 1 x5 + 2 = 7; 28 is divisible by 7

Divisibility by 13

Ekadhika for 13 is derived from 13 x3 = 39 and therefore it is 4

13; 3 x4 + 1 = 13

39; 9 x 4 + 3 = 39

SUB SUTRA 6
YAVADUNAM TAVADUNAM

i) In two general equation such as, ax + by = p and cx + d y = q, where x and y are unknown values, then by solving them simultaneous equation we can get the values of x and y

x = (b q – p d) / (b c – ad)

y = (c p – a q) / (b c – ad)

We can use the above formula directly;

Examples:

1)4x -3 y = 2

2x + y = 3

x = (-9-2)/ (-6-4) = 11/10

y = (4-12)/ (-6-4) = 8/10

2)3x – 5y = 3

2x – 3y = 4

x = (-5x 4 – 2x1)/ (-3x2 - 3 x (-3)) = -22/3

y= (2x2 -3 x (-3))/ (-3x2 - 3 x (-3)) = 13/3

(We can use elimination, substitution or cross multiplication method to solve the equation in conventional method)

ii) Used for finding the cube of a number that is close to the powers of 10 such as 10, 100, 1000...etc.

Example:

997^3

Base = 1000; initial deficit = - 3.

i) Number + twice x deficit= 997+2 (-3) = 991

ii) New deficit x old deficit= -9 x -3= 027

iii) Cube of initial deficit= $(-3)^3$= -027

$997^3 = (997 + 2 \times (-3)) / (-9 \times -3) / (-3)^3$

= 991 / = 027 / = -027

= 991 / 026 / 1000 - 027

= 991 / 026 / 973

= 991026973

SUB SUTRA 7
YAVADUNAM TAVADUNIKRTYA VARGANCA YOJAYET

I) It is applicable to find the square of a number close to bases of power of 10

Here, whatever the deficiency subtract or add that deficit

from/to the number and write alongside the square of that deficit.

$8^2 = 8\text{-}2/2^2 = 64$

$97^2 = 97\text{-}3/\ 3^2 = 9409$

$14^2 = 14\text{+}4\ /4^2 = 1816$

$102^2 = 102\text{+}2\ /2^2 = 10404$

II) It is applicable to find the cube of a number close to bases of power of 10

Step 1: number + deficient+ deficient = left side

Step 2: (number obtained in step 1 – base) x deficient= middle number

Step 3: cube of the deficient = right side

$102^3 = (102+2+2)/\ (106-100) \times 2/\ 2^3$

= 106/12/08

= 1061208

98^3 = (98 -2 -2) / (94-100) x -2 / $(-2)^3$

= 94/12/-08

=94/12-1/100-8

= 94/11/92

= 941192

SUB SUTRA 8
ANTYAYORDASAKE'PI

The Sutra signifies numbers of which the last digits added up give 10.

The Sutra works in the multiplication of numbers in which the sum of the last digit of the first number to the last digit of the second number is 10. Further, the portion of digits ornumbers leftwards to the last digits remains the same.

Step 1: Use Ekadhikena on left-hand side digits.

Step 2: Multiplication of the last digits gives the right-hand part of the answer.

Examples:

62 x 68 = 6 x7 /2x8= 4216

265 x 265 = 26 x 27 /5 x5 = 70225

111 x 119 = 11 x 12 /1x9 = 13209

84 x 86 = 8 x 9 /4x6 = 7224

92 x 98 = 9 x 10 / 2x 8 = 9016

SUB SUTRA 9
ANTYAYOREVA

'Only the last terms'

This is useful in solving simple equations of the following type.

I) $\frac{AC+D}{BC+E}=\frac{A}{B}$ then $\frac{A}{B}=\frac{D}{E}$

$$\frac{x(x+2)+1}{x(x+3)+4}=\frac{x+2}{x+3}$$

Therefore, $\frac{x+2}{x+3}=\frac{1}{4}$ solving x= -5/3

II) $\frac{1}{(x+3)(x+4)}+\frac{1}{(x+4)(x+5)}+\frac{1}{(x+5)(x+6)}$

$$=\frac{3}{(x+3)(x+6)}$$

The sum of the series is a fraction whose numerator is the sum of the numerators in the series and whose denominators is the product of the 2 ends (binomials).

SUB SUTRA 10
SAMUCCAYAGUNITAH

"Gunitasamuccaya"

To verify the result, the product of the sum of the coefficients of in the factors is equal to the sum of the coefficients in the product.

Examples:

$2x^2 - 3x - 20 = (x-4)\ (2x+5)$

Or, $2-3-20 = (1-4)\ (2+5)$

Or, $-21= -3 \times 7$

Or, $-21 = -21$

$x^2 - 5x + 6 = (x-3)\ (x-2)$

Or, $1-5+6 = (1-3)\ (1-2)$

Or, $2= -2 \times -1$

Or, $2 = 2$

SUB SUTRA 11
LOPANASTHAPANABHYAM

"By alternate elimination and retention"

I) We can apply this sutra to find HCF in easy way

$x^2 + 2x - 8$ and $x^2 - 6x + 8$

Consider,

$x^2 + 2x - 8$

$x^2 - 6x + 8$

$\underline{- \quad + \quad -}$

$8x - 16 = 8(x-2)$

HCF = (x-2)

Cross checking by factorization method:

$x^2 + 2x - 8 = (x-2)(x+4)$

$x^2 - 6x + 8 = (x-2)(x-4)$

Therefore, HCF = (x-2)

II) This sutra is used to factorize complex equation by eliminating any of the variable

Factorize $2x^2 + 8y^2 + 5z^2 + 8xy + 8yz + 7zx$

We have 3 variables x, y, z.

Remove any of the variable like z by putting z=0.

Hence the given expression

$E = 2x^2 + 8y^2 + 8xy = (x+2y)(2x+4y)$

Similarly, if y=0, then

$E = 2x^2 + 5z^2 + 7zx = (x + z)(2x+5z)$

Final factors can be given as

$E = (x+2y+z)(2x+4y+5z)$

SUB SUTRA 12
VILOKANAM

"Solving by observation method"

There are instances when we come across problems that can be solved by mere observation. But the hint behind the Sutra enables us to observe the problem completely and find the pattern and finally solve the problem by just observation.

Examples:

$x + 1/x = 10/3$

$x + 1/x = 3 + 1/3$

Hence, $x = 3$ and $1/x = 1/3$

$y + 5/y = 21/4$

$y + 5/y = 4 + \frac{1}{4}$

Hence, $y = 4$, $1/y = \frac{1}{4}$

SUB SUTRA 13
GUNITASAMUCCAYAH
SAMUCCAYAGUNITAH

To verify the result, the product of the sum of the coefficients of in the factors is equal to the sum of the coefficients in the product.

Examples:

$4x^2 + 12x + 5 = (2x + 1)(2x + 5)$

Or, $4 + 12 + 5 = (2 + 1)(2 + 5)$

Or, $21 = 3 \times 7$

Or, $21 = 21$

$2x^2 + 7x + 6 = (x + 2)(2x + 3)$

Or, $2 + 7 + 6 = (1 + 2)(2 + 3)$

Or, $15 = 3 \times 5$

Or, $15 = 15$

SPECIAL USE OF SUTRAS IN DIFFERENT CONTEXT

1. PARTIAL FRACTIONS

Paravartya Sutra can be used in partial fraction problems in Integral calculus.

Example:

$$\frac{3x+5}{(x+1)(x+3)} = \frac{A}{(x+1)} + \frac{B}{(x+3)}$$

We can find the values of A and B by using Paravatya sutra:

Step 1: To find the value of A, we equate its denominator to zero and thus get x = -1, then substitute the value x= -1 in the LHS, except in the factor which is A’s denominator.

Step 2: To find the value of B, we equate its denominator to zero and thus get x = -3, then substitute the value x= -3 in the LHS, except in the factor which is B’s denominator.

$$A = \frac{3(-1)+5}{(-1+3)} = 1$$

$$B = \frac{2(-3)+5}{(-3+1)} = 1/2$$

$$\frac{3x+5}{(x+1)(x+3)} - \frac{1}{(x+1)} + \frac{1/2}{(x+3)}$$

So, we can find the values of A and B mentally and verify the result which we get by using conventional method.

2. ALL FROM 9 AND THE LAST FROM 10 (NIKHILAM SUTRA)

Applying this sutra to any number gives us the deficiency of the number from its base (nearest power of 10)

Example: 674

By using Sutra "all from 9 and the last from 10 "the deficiency of 674 from the base 1000 = 9-6/ 9-7/10-4 = 326

3. BAR NUMBERS (VINCULUM)

$1\bar{1} = 10 - 1 = 9$

$6\bar{3} = 60 - 3 = 57$

Subtraction using vinculum makes the process very easy

Examples:

 569
- 383

$2\bar{2}6 = 200 - 20 + 6 = 186$ Or $2\bar{2}6$ = 2-1/10-2/6= 186

 58936
- 39087

$2\bar{1}9\bar{5}\bar{1} = 20000 - 1000 + 900 - 50 - 1 = 19849$

Or $2\bar{1}9\bar{5}\bar{1}$ = 2-1/ 10-1/9-1/10-5/$\bar{1}$ = 1/9/8/5-1/10-1 = 19849

By regular practice it becomes very easy to find the result mentally.

4. MULTIPLYING THREE NUMBERS TOGETHER SIMULTANEOUSLY (NEAR TO A COMMON BASE) USING NIKHILAM SUTRA

95 x 97 x 98

Step 1: Find the deficiencies. Take one number and subtract the other two deficiencies from it (first part of the answer)

Step 2: Multiply the deficiencies together in pairs and add the results (middle part of the answer)

Step 3: Multiply the deficiencies together (last part of the answer)

Base = 100, deficiencies= -5,-3 and -2

95 x 97 x98 = 90 / (2x3+ 3x5 + 5x2/ (-5 x-3 x-2)

= 90/ 31/ (-30)

= 90/ 31-1/ 100-30= 903070

5. DIGIT SUMS (DS)

The Digit Sum of a number is found by adding the digits in a number and adding again if necessary until a single number is obtained. If any number consists of 9 in between it can be cast out while adding.

DS can be used for cross-checking multiplication, addition, and subtraction (special cases for subtraction) results

Example:

Multiplication

233 x 3 = 699

DS 233 = 8; 8 x3 = 24; DS = 6

DS 699 = 24; 2+4 = 6

Addition

  567
+ 253

$7/11/_{1}0$ = 7/11+1/0= $7/_{1}2/0$ = 820
DS 567 = 18; 1+8 = 9
DS 253 = 10; 1+0 = 1

10; 1+0 = 1
DS 820 = 10; 1+0 = 1

Subtraction

558 ----- DS 18; 1+8 = 9
-341 ------ DS 8 so, DS 9-8=1
----
217 ------ DS 10; 1+0 = 1

6. DIVISION BY POLYNOMIALS

Example: $12x^2 - 8x - 32 \div x-2$

STEP 1: $12\,x^2/x = 12x$ (part of Q)

STEP 2: 12 X -2 = -24; reverse the sign = 24. Then add to the next coefficient i.e. -8; 24-8= 16 (next part of Q)

STEP 3: Multiply 16 by -2 and reverse the sign we get 32. Add 32 to the next term in dividend; 32 -32 = 0 (R)

Quotient = 12x +16   Remainder= 0

SUMMARY

SUTRA	NAME	MEANING	APPLICATION
Sutra 1	*Ekadhikena Purvena*	By one more than the previous one.	Squaring numbers ending in 5 Calculating the value of 1/19,1/29,1/49 etc. by using division method
Sutra 2	*Nikhilam Navatascara mam Dasatah*	All from 9 and the last from 10	Multiplication of numbers close to 10, 100, 1000...etc. Squaring technique Division by 9
Sutra 3	*Urdhva Tiryagbhyam*	Vertically and Crosswise	Multiplication of large numbers
Sutra 4	*Paraavartya Yojayet*	Transpose and Apply	Division of numbers when divisors are slightly greater than 10, 100, 1000... etc.
Sutra 5	*Sunyam Saamya Samuccaye*	When the sum is same, it is equated to Zero	Used in different places, where sum is equated to 0
Sutra 6	*Anurupye - Sunyamanyat*	If one is in ratio, the other one is zero.	Solving simultaneous linear equations
Sutra 7	*Sankalana - Vyavakalanab hyam*	By addition and by subtraction.	Solving simultaneous equations where x and y coefficients are interchanged.
Sutra 8	*Puranapuran abhyam*	By completion or by non-completion.	Finding the roots of the equations.
Sutra 9	*Calana - Kalanabhyam*	Difference and similarities	Finding roots of equation

			Used in differential calculus
Sutra 10	*Yavadunam*	The deficiency	Applicable to obtain square of a number close to bases of powers of 10
Sutra 11	*Vyastisa mastih*	Part and whole	Solve the equation, by finding factors
Sutra 12	*Sesanyan Kena Caramena*	Remainder by the last digit.	It is to express a fraction as a decimal to all its decimal places
Sutra 13	*Sopantyadvay amantyam*	Ultimate and twice the penultimate.	Solving equations
Sutra 14	*Ekanyunena Purvena*	One less than the previous one.	This sutra is used in case of multiplication by 9, 99...
Sutra 15	*Gunitasamuc cayah*	The product of the sum is equal to the sum of the product	Used to verify the correctness of obtained answers in multiplications, divisions and factorizations.
Sutra 16	*Gunaka Samuccayah*	The factors of the sum are equal to the sum of the factors	Cross checking factors
Sub-Sutra 1	*Anurupyena*	Proportionately	Finding the product of two numbers that are close to common base powers of 10 (10 / 100/ 1000...etc.)
Sub-Sutra 2	*Sisyate Sesasamjnah*	Remainder remains constant	A corollary of Nikhilam Navatasccaramam Dasatah.
Sub-Sutra 3	*Adyamadyen antya-*	The first by the first and	Factorization of quadratic equations

	Mantyena	the last by the last.	
Sub-Sutra 4	*Kevalaih Saptakam Gunyat*	In case of 7 multiplicands should be 143.	Division based on 1/7 = 0.142857 (approx. 0.143)
Sub-Sutra 5	*Vestanam*	By osculation.	Perform divisibility checks for numbers.
Sub-Sutra 6	*Yavadunam Tavadunam*	Lessen by the deficiency.	Solving equations, Finding the cube of a number that is close to the powers of 10 such as 10, 100, 1000...etc.
Sub-Sutra 7	*Yavadunam Tavadunikrty a Varganca Yojayet*	Whatever the deficiency, subtract that deficit from the number and write alongside the square of that deficit	Finding squares/cubes of numbers close to 10, 100, 1000...etc.
Sub-Sutra 8	*Antyayor Dasakepi*	Sum of last digits is 10.	Multiplication of numbers whose last digits add up to 10, 100. 1000... etc. if the remaining digits are the same
Sub-Sutra 9	*Antyayoreva*	Only the last terms.	Solving special types of algebraic equations.
Sub-Sutra 10	*Samuccayagu nitah*	By the deficiency.	Used in special cases.
Sub-Sutra 11	*Lopana Sthapanabhy am*	By alternate elimination and retention.	Finding HCF of algebraic equations.
Sub-Sutra 12	*Vilokanam*	Observation.	Solving problems by making simple observations. Can be applied in: Partial Fractions

			Simultaneous Linear Equations Simultaneous Quadratic Equations
Sub-Sutra 13	*Gunita Samuccayah-Samuccaya Gunitah*	The product of the sum of coefficients is equal to the sum of coefficients in the product.	Factorization and solving of algebraic equations.

REFERENCES AND ACKNOWLEDGEMENT

Vedic Mathematics, Jagadguru Swami Sri Bharati Krsna Tirthaji Maharaja

Website - mathslearners.com

Website - collegedekho.com

Website - tiwaryacademy.com

I am thankful for various other books and websites also, for helping me to bring out this handbook. All the references are used for educational purposes.